MW01428942

charger

Also by Margaret Christakos

Charisma
Excessive Love Prostheses
Her Paraphernalia: On Motherlines, Sex/Blood/Loss & Selfies
The Moment Coming
Multitudes
Not Egypt
Other Words for Grace
Sooner
Space Between Her Lips: The Poetry of Margaret Christakos
Welling
What Stirs
Wipe Under a Love

charger

for Shawn

poems

Enjoy!

MARGARET CHRISTAKOS

love, Margaret

12 July 2020

Talonbooks

© 2020 Margaret Christakos

All rights reserved. No part of this book may be reproduced, stored in a retrieval system, or transmitted, in any form or by any means, without the prior written consent of the publisher or a licence from Access Copyright (The Canadian Copyright Licensing Agency). For a copyright licence, visit accesscopyright.ca or call toll-free 1-800-893-5777.

Talonbooks
9259 Shaughnessy Street, Vancouver, British Columbia, Canada V6P 6R4
talonbooks.com

Talonbooks is located on xʷməθkʷəy̓əm, Sḵwx̱wú7mesh, and səl̓ilwətaʔɬ Lands.

First printing: 2020

Typeset in Minion
Printed and bound in Canada on 100% post-consumer recycled paper

Interior and cover design by Typesmith
Cover image: mosaic charger, February 29, 2016, from *#chargerseries* by Margaret Christakos

Talonbooks acknowledges the financial support of the Canada Council for the Arts, the Government of Canada through the Canada Book Fund, and the Province of British Columbia through the British Columbia Arts Council and the Book Publishing Tax Credit.

Canada | Canada Council for the Arts | Conseil des arts du Canada | British Columbia | British Columbia Arts Council An agency of the Province of British Columbia

LIBRARY AND ARCHIVES CANADA CATALOGUING IN PUBLICATION

Title: Charger : poems / Margaret Christakos
Names: Christakos, Margaret, author.
Identifiers: Canadiana 20190194774 | ISBN 9781772012491 (softcover)
Classification: LCC PS8555 H675 C43 2020 | DDC C811/.54,—dc23

for my children

and I caress you now with the same touch
as I caress these keys.

—JORIE GRAHAM
in *Fast* (2017)

because all of us are moving
& may detonate the earth at any moment

—ERÍN MOURE
in *Search Procedures* (1996)

1

passing on information

something has to happen

for something to occur in memory

for something to touch my insides

information moves

and becomes

memory occurs

feel free

feel free to get

feel free to

go ahead and make

say i never told you

does not make us remember it

inside the organism for memory to occur

the skin is a membrane

i have to involve it in movement

when it seeps through time

material

in the dwell or the repercussion

to argue

stuck in / on / over this

make a big deal of this

a mountain of this molehill

or it went right by in a blur

say you missed it

about who won all the awards who should've won them

tug at my bra strap and say

pop by in a couple days   say

and it went past me   i let it go past me

like a wind made

i didn't feel it

what day is this we're   talking about

go through my inbox i'll

i'm interested in reading

and i am a good swimmer

i am   turning

i am skilled at yanking the water past my body

in the cascade of posts yesterday

for their better excellence

listen you're not listening

oh my cell died i forgot my cord

felt it slide over me

of data

or if i did i don't remember

let me

get back to you promise

the article you sent

i hold my breath

blue

and jiggering forward

like some large lure waiting

if the great mouth opened

its hot insides

i felt like a gulp of bad news except i was the part that

i was in the inner room where the real decisions

do you know

i'm telling you about this later

it

somebody please write something

add me to your cenotaph

where something like golden blood

things are seeping

it's not pretty is it no

for a mammoth fish to rise

i would feel myself pulled into

the larynx like goo

couldn't be spewed on the sidewalk anymore

move on to the next level

what i'm saying

beyond the unavoidable

happened

and submit it to a magazine

scratch our names into a wooden trunk

seeps from the tree's subdermal pulp

everywhere we look actually

it's not pretty

when i check in that mirror

there's an abrasive noise hitting

i'm getting whacked like a new form

but that somebody

for everything it

just slid past and

look over your shoulder

who really wanted her to follow you

come on

we say to appear equanimous

than marching

desert

a knapsack of

i can see my hair's wrong

one side of my head over and over

of art that nobody likes yet

somewhere   will recognize

has to offer the moment we

forgot to remember

so she disappears again

all the way to the surface?

that's just something

there's nothing sweeter

toward a scorching

with bronzed shoulders and

steel weapons

the surveillance cameras

don't think you're

when the wind gathers

pull some cotton

imagine it's made of language

outside the head

with a new efficiency

i'm passing you on this message and

this is pure

this is

i think you should

will catch it

out of danger

it's clearly full of information

over your body

which used to go on

but now swirls from gaze to gaze

don't think it's casual

steam

something else

read it

2

what

ready to roam

you are ready for a decent

you are good to go

you are   is   fully charged

amount   of   time

good

to   go

fulla   juice

look at that   white   bar

beautiful   thing

sorry   is that a   plug   there?

—there—

between your   skin   and

yes   it's a   plug

we   are   good to   go

ten   fifteen   should be

where

the wall yes there

of course please

no problem —there—

great

great awesome yes thanks

sure no

what charges you up

down

up

how the spark gets

how

leap

void

why

how usage

hold

shush it in a

go over to your

definitely it's

what changes you

to up

to down

in the plug

the sparks

the

we speak

peaks

that thought

while

mother's grave: *what?*

go cover

space her

brief pause

all of her

go

over hers

observing

on its

out of

a core

your mother's

on the planet

after-effects

hold your body

like a drone

what's

path

the past toward

source

now

she's possible:

i am

for the little *ding*

ready

your legs

be rude but

lemme just

she was—

really?

listening

that means

my charger's

between

don't wanna

—there—

good

up

i love

that white

solid

because?

have so much

all charged

—awesome—

when

bar's all

again

because

time i'll

again

the time spent

aglow

—yeah like a ghost—

zombie city

rolling with

soul

time

i'll double it

against

dead on the table un-

aghast

oh here we go

to hauntsville

misty hills

discombobuli lost

central

what?

i was just

for my charger

one who roams

a vague

a roadless

worrier

or else

to rove to

aimfully

waiting

to charge

some say

is a vagrant

set of regrets

to roam is

ramble

surf

amped up

for a few hours

on the

loose

chill

meaning

ready

for the beautiful

to go

decently

s'good all's

recharged and

i'm ready

thing

now

do you mean—

unplug

i'm

souvenirs

with the

and i love

like

solid as

vision all my

yes let's

me

fulla

filled

white juice

that feeling

bam

lifted light raised

knowing incandescent

you know what i

really

like?

the plug?

a nice clean

with nothing

and i can just

wait you like

finding it—

yeah the outlet

fresh outlet

in front of it

stick it in—

to wait?

well

i wouldn't say

i hate waiting

it's like being

some kinda

religion of

or making an onset into a

fully

on

actually makes my mouth

that—

wasting the sap

of my time

and unbeing

debt in

solitary

solid white horizon

juiced

salt water waiting

water

that's

strange   since

makes

turn   to

right   but

don't   be   jumpy

making light   like

the   rest   of these

possible

waiting

my spit

wine or

milk oh

you're *special—*

i'm just sitting here

chargés d'affaires becoming

again

no   problem   —you're

wi-fi

access for   everyone   the

all   need

to   roam   free   and

so

replenished

welcome—   it's

a   free   country

equal

normal   after-effects

you   know we

more

juice

solid

what?—

so   ready

to   go

3

go to binary categories

pump out a little

just also

listening

wind

like a shore

with a herd of cattle

fresh

and fuck with them

tertiary ambiguity

sound out

to that

hiss

on fire

thirsty for

grass

more like   colville's   black stallion

obsolescent

god   bless the

trespassing

zero

in her

it's dusk   she

she   doesn't think   about

more

airborne and   charging   an

train

pissed-off   skateboarder

the track

devices

backpack

couldn't   care less

you any-

let

her   float

offline

thrust forward

chaos's

chasm

who you

after so politely

to respect your

do you bet

be

there

flips

on its side

charges the

into memory measuring

gonna call

asking them all

privacy who

will

when the fence

and a flock of crows

bandstand?

i commissioned

from the friend

would make it

sometimes

to   account

glitches   like

ghosting

a smoke-beige

by a simple

the gold smell of

a fancy analogue painting

i always knew

holding a dear one

an ignited rectangle

screen pierced

nailstem and

sulphur

whispered the cheshire officer

busy

catching some

i don't know anyone

. than me

think

wind today—

it

drink me deep

her fellow troops

virtual *zzzz*'s

who'd be faster

posting

what d'you

of this

isn't

eerie?

4

what we had to   cover

toward a stylish   apology

to be sincere   yet   also

shaking

a stylish

a mouthful of   wan words

disembowelled fish eyes

at a warm bowl of

goes a long way

one that   seems

never falls on its   haunches

apology

manages

as if they're

looking   frankly

blood

get up

chess

swagger your

shame-giggle

you deserve

observe

how many pix of her

how many intentional brownouts

operate like a new car

on the

board

in a loud electrobeat

awake you

others

did you instagram

can you

impressing the relatives

go into a quiet room

overcome all the circuits

get outside of

there's no neighbour

anything

just fill it

we're sure are tracking the true winner

a cup of sugar

slow-baking

honestly

it isn't coming out

that's hard

when crowds form

notice— you

go charge into

power

but only

you can unplug yourself from

it's

suspressing truth—

for the cause you embody:

are protected—

your circuit of

and light—

if

the damn cool of

the shade

i heard it coming

i heard it

or snore

head somewhere kind

turn a clear profit or—

why didn't you file

a warning shot

in the dream remake

the nearest dollar store maybe

and

what's stopping you?

5

can i cease and

i'm right

remembering when

just

and that day

in a closet

as if the rush

door

would evaporate

attention

desist you yet?

here like a socket in reverse

you were

a baby

we hid

cowering

of water under the front

if left to its own

span

6

woke up

about

the weather report for

referred to as .

into the hospital room

you were gone

shiny green patches where

would be like

hallucinatoria

thinking

laudanum then heard

southern ontario

weird

two steps

i knew

even if i'd brought proof

adolescent mallards

have

their ears

metal or dreams of

honked   in a quiet

    a lot of   the bad

with recognizing   another   shudder

    alongside

                                            our little

    after   bunches

                                            stop

quawk *qwauk*

tone

crowds out

thoughts

we live

in the sky

rainbow tinkle-fest

of thunder

drop and

roll

contented

rigor mortis

scale saps

to the orderly who

her shift

no tip

ha-ha

get on with it

clawing

overhead

maybe begging

on the richter

our wide-irised

okay

lets us know

is closing soon

just could we

your hands

at the light

for the bag

your palms

actually your knuckles

calcite-thickened

ours the firm

forearms

your shoulders

common give-it-up

are

were

ours

ours

your

nails

wrists and soft-furred

ours

jiggle

s'just a

reflex ma'am

absurdly

i was reading aloud

alternated

spooning you

several

in me

opening

teenaged

the part where marian

what the hell ay

before the   walk and the   ducks

atwood's   *edible woman*

with

some final   vanilla ice   cream

mouths

closing   and

like   a   silent

gut-cry

shimmies   down under the   couch

yes   what the   hell

because even

said anything as crass

you raised us

the moment

acceptable

to have

upper

hand him over when we're

ready

just

measure become as indelible

as   you left   i wouldn't've

as   what   the   eff

well   and now's

to show   it's

for   heaven

the

hand

we'll

good   and

ten   fifteen   sure

trying   to   let   this

as possible   thanks

what's this

i've always had

yours

shadows hands

make

against my bedroom

always anew utterly

mid-afternoon light on

of young ducks every

of the thumbs'

obsession

with the hands

my own

my children's—

dances they can finger

ceiling ways they touch

balletic expressive as

the narrow skull-muscles

hue gesticulating a semaphore

belligerent lingo

in a minute

this public-share memory

the social

equivalent of a

of ice cream—

your cellphone

your hands bright-naked

stroking

would you bet

atwood was

under the couch

even letting her make

for the

when you feel

has merit as

media–post

spoonful

discharge

to render

and chilled as if

rain from a mallard's bare neck—

your last silver dollar

gonna leave marian

forever not

a jokey *quawk* *qwack*

win?

7

among poplars

for the wires

innards with

my own greedy

all the anecdotes

in case

really though i am leaning

the natural world's

as if it exists

because there are trees

whether we've rooted

our relationship

i am   looking

that feed   this house's

light—   i am   measuring

ignorance   and

i could   switch on

of   an outage—

out   toward

blizzard

wherever   there   are   trees

wondering   as static snows

enough   to   hold   ground—

despite the distance   are there

into   your   appetite's

when all   that's

grove   of wintering

ways   we   text

to   fuck   bodies   beyond

the   online   dating

a   burn   at matches

flawed   and frosted out

hot-wire   sex   on an   air

with descant   potty mouth

to a   brothel's

signals   that feed   me

innards   even

obvious   is a

trees   and the   fast-tapped

each other   persuasions

our   own—   yet i’ve dumped

site   in a bit   of

that seem   well

plus i am   humbled   how we

mattress   in the   cellar

the neighbours   could chalk up

successful   home op

still we're

hobbled in

not saying enough

amid squalls of affection

as they are

these poplars are tethered

imbroglio while

exhale

virtual frame of

street lights

lamp bulbs

their timers'

middle-earth's facetime

about our particular wounds

as diligently casual

intimate—

i'm weighing the ways

by a gnarled root-

their crowns

freely into a

sleeting sky

the

are coming on

pulsate

efficiencies and at one stroke

i'm engulfed in

unable

about monogamy

what i'd wish of either

in them enough and our fondness

recent

or photo albums

legitimacy—

oscillating trunk

its sizzle of ice

electric   society

to darkly   mope

and fidelity   or

if i believed

is   relatively

without   seasons

or family-woven

just   our   test-pattern   vertigo

against trunk

and volt and

melt

8

if you are

what's the

of plugging back into

with all

seriously

to be retrieved

handful

one's warm

explain it to our

made this special trip

sunday no

all of us

about to die

point

that massive panel

the tubes and

the hissing

what's

in another

of your loved

flesh please

assessors who've

in on a

less we're

ears

9

beyond the windowed

are encasements cardinals

in synchrony

against the tilted

and white

what of it all

strums my face with

fevered looking—

this

specific moment

attention—

well

rectangles we know

seem to be chirping

which clatters

planes of blue

and sun-illuminated air—

is a good song what of it

a human wish —my

hope to inhabit one

an increment of

all of it

presumably

this   is a

smell

beyond   glass

that seems to

by   these   bare feet

by   these

skull-case

away from

toward

as it   is

as it

meadow without

amphitheatre

air outside

charge interior—

i know myself

tittered plucks in my

by my fidgeting

utopia

something as ugly

particular

as off-key

is detectable

like a cadre of

my mother's

a living room with

paintings and rescued

and plenty of

caved-in

that vague

centrifugal mirage of the

on sunday morning sinatra and roberta flack

the bodies of your

memories   of

        way   of   doing up

        a dozen   artists'

        driftwood   stumps

ashtrays   near   the most

cushioned chairs

        cloud of   smoke

vinyl   she liked   to play

        instead of church   or garden work

                you   miss

        parents   so intensely

that a window

in the past left for

and ed sullivan

moments of the really

be their fifteen

embodiment entering

right all rigid

the honoured guests

that lazarus

the commoners after

serves a purpose

priests

framing his magic

big reveal

it could

minutes of recuperated

from stage left or

in discomfort at being

in the funhouse

likes to open to

the ratings season relaxes

offstage at

of the old chesterfield

and the *globe* between

mugs of coffee

experiment in

how they

television

updates as the

between them

of mutual

you can't suckle

imagine

but both   preferred   to stay

either end

downstairs   *northern life*

cooling off   like a   biology

evolutionary   psychology

acceded   to

and   local   political

two-way   mirror

glinted   their   decades

witness

in what   you

of anyone   else's   romance

the top 40

all of a sudden

with the

convergences

their

with your own

frame

nibbling back the

momentary

you don't define

springing to cue

tittering the ears

livid

others call

lives

you just persist

willingness to walk into

hear birds

skin of this

retrieval

this

winged idea of the

singing our lives

resurgent tune’s

stranger dimension

on

replay

10

if

alive sperm

live egg to

the reverend

argues

surely life

to begin

famous fetal spark

doesn't

life is a

overarching

miracle of

it takes

and a

ignite

conception

abortionist

then

cannot be said

with that

of fertilization

follow

continuous condition

the individual

a particular pregnancy

and so

her own right of

she is the

and may

offspring's ticket to

ready and

the better arguments

for respecting

the mother over

the zygot

on this greyish

a woman bears

choice immaculate

livid subject

call down her

ride when

good

one of

i’ve heard

the personhood of

a personhood of

thanks cbc

sunday afternoon

as i sweep

and sip

sweet wine toasting

liberties

depleted plumpness

pair of lips

in this house alone

thin dress and

achieving both

communion with the

of public

make eggs or

with myself

checklist of

the   kitchen   floor

an   early   glass   of

     my   menopausal

and   pinching   the

     of my   lower

          since   here

     i   can   wear a

nothing   else   while

     housework   and

               social   prosthesis

     radio   i can

               make   hay

mastering   a quick

          personal   redemption

knowing i

defer two of

until i could

is to divide

in one’s galaxy

trompe l’oeil

ground flickering

the naked mind

how the

ever happens

clipping one

its vessel

squeezing

chose properly to

you for later

handle how to mother

all the cells

into a two-way

the figure and

continuously so rapidly

cannot fathom

artful switcheroo

without a hiccup

vesicle from

or the iris

its lid against

the oncoming blade

the thought

the act

both are not

their own

with

or wine

or

you get

at any

it

baby to slip

of cosmic life

the sudden

unsplicing the act and

about

as if

always refilling

bulbous glass aglow

milk or

or

blood

the ideas

rate

takes two

over the high reserve

force into

breath-jolt

of one plug-in into

that is always also

toward a more thorough mapping

pre- conscious about the infinite

mad dash

tree trunk which each

gravity's omnipresent

smashed

not-so- nimble

to ably demonstrate

final it all

insurmountable

not rising

as if

this   power way station

springing   forward

of the   universe—

litter   of young   squirrels   in a

for   the   nearest

scales   against

thirst   for   the

spread-eaglet   of   the

squirrel-pup   in its   fall

death   how

is   how

as if   wisps   are

into the   stratosphere

tendrils

do not

we fantasize

inverse of all

silent with our

followed by that official

whistling for us

come into

come soft

this

insinuate all

into the trippy

we mourn

gentle word arrangements

one o'clock time signal

chorally——

my arms

darlings

into

Cloud

11

slides

over

charges us—

i could have interrupted

the thing that

every other

sounding

that afternoon

and demanded

peace— someone nuptial

before vowing 1993

my heart's

didn't— let it all

imagine the scowl on

this

our rose garden party

a justice of the

in official garb—

to always let you know

aspirations

but

just bloom

my face as i write

now

with you poem-bombing

what is supposed

other

my parents' final words 2011 / 2015 our

the sun's partial eclipse

in the right place at

to wreak

being miserable stewing

an ovoid grey lake

under this dock leavening

steady glamour

living— living—

to unsnarl several

related losses—

son's friend's last breaths 2010

by a greedy moon almost

the right time ready

maximum drama 2017 and everybody

in memory which is

lashing at the shore

my heart through its

of living— living—

living— wonder of being

fully   juiced.

we

sense the sounding

wind

at our

luxurious exhalations

keep

inhale sliding

of a suede day's

pushing

inhalations then slow

and repeat—

repeating— keep the

soft over the

exhale—

and repeating

even if twisting his

waist in the suburban

at dairy queen made

him pant and pant 2009

and there was no

camouflaging the copd

anymore from

the kids

what we know about

being alive when we

are alive is like

a lunch gong going

off 1968–1978 in the canyon—

get our butts

home now because

it’s ready for us

she gets my email 1998

as our small feet

hit the black rock 1971

i 3D-print them 2019

grilled cheese open- face

sandwiches 1973

we   facebook-send a photo 2007

with   the   chocolate cream   cake

from   johnny's   i   know

he's   coming 1976

the   black   burnt   film

on   top   —pure

carbon—   alongside   her

coral   lipstick   tube   on the   counter 1972

they set spotify 2015 to

we are the champions 1978

the steely grey-green sky

when the winds hit when

scrambling to the basement

long enough to blindside

let you live-stream

the damn eclipse tell you

one more time about

that morning 20 aug 1970

the hurricane sent us

when sunlight disappeared

roofs and fizzle power

there's a spaceship

still convincing with

can we admit

such a vessel remains

generations'

a future?

my north shore heart's

includes

our children's shared

across the

you accompanied me through

on   ramsey's   southwest   end

its   surcharge   of   future-vision—

now   how   radical

for   forthcoming

belief   there   will   become

fuel

this   view—   you and

time   travels

twenty-odd   decent   years

the   motions

i wish you well 2026

but not again in the middle camp's flickering

black-eyed susan grove

with our

amidst

all that charged us—

over the exhale

will see you in some afterworld

birch shade near the decapitated

a half turn from the ancient crabapple

broken Common Law

the inhale of

our slow slide

still here

12

it's like

powered

mower arrives

duty—

the machine operate

puncture

as nature's score

and licks

breeze

isn't it

a solar-

lawn-

to do his

both the guy and

their powers to

what we refer to

of laps

and skirmishes of

and christ but

agitational?

i just

a little

of them actually out of a van

chorus of

over the short

making them

green more

down to echo

ground itself

wanted

quiet—   now   three

off a trailer   a   coordinated

mowers   moving

green   expanses

shorter   and   less

uniform   more   ground

the   contours   of the

and   i   know

it’s their summer

students at the

two weeks’ final

fall classes

at the very

footing their own

instant

liking and

when their palms

job   and   they're likely

   college with

pay   before

             begin   and

least   they're

smartphone   bills   so

   texting

swiping

             will   continue

are   ready   to   charge   again

primed to guzzle

the shared social

and what's

natural about

take this

open

place and

some

it's

sound

upon

all-we-can-eat

so special or

quiet anyhow?

wind- scuttled

concept

set it in

walls—

all

never stillness

if we

to john cage

beyond

every

in the

gets

to

that is bearable—

twitch—

pay attention

if we hear

the measure—

body

room

choose an interval

our hands

fold inward—

quickens

more

less

like a cow

stomachs

we're

for

to be

our arms

our brain

for something

intense or

streamlined

with four

restless

all the chambers

filled in

with

juice

and make—

and

it's a

pre-millenial

for all of us

our bodies

so i get when i'm teaching

at break

students

in their

the white

we both milk

drink

emit—

choice

generations took

an urge

bristle to bear—

why

all my

stay

seats wordless

and feeding

streams of

sneak

a quick

red flare

on the text message

because

to gorge

french kisses

is killing

with its

unisonic

on the slick

digital updates

while i

look

for a

alit

icon

later i want

on your

and the suspense

me softly

song-like

pantomime

a set of   echoes   all the   bodies

can't help   but   quaver

like a   dun-green   field

at   sundown   moaning

for   farmers

at chest level   for   us—

who   are   being

the

in this chamber

together

of cows lowing

to load the feed

we

so disciplined about

interval

gertrude stein told us

a sentence

how
a thing

knows itself

to be

as a

of its own

the multiple other

waiting obediently for the

the infill to burst

was how

actually happens

to be filled in enough

as discernible

stomach

alongside

stomachs

brain-spill to arrive

aplenty into

radiant data bins

we love

be

they'll

full again

futurity's poetry

of the

all the new

by three solar-

machines

their iron trailer

some

to believe will

obviously refilled—

how

be

of

news

news smooth-chewed

powered mowing

now corded up on

drenched in

later descendant sun—

a caravan   led by

hand   smuggling   a cellphone

mirror

lax hand   steers past

on a field near

toward a   neon

dusk-brightened   all-beef-burger

with   ms   lauryn hill

it's still

one young tanned

below the rearview

while his second

a lowing herd loose

the emptying college

evening class

drive-thru

on the van radio like

the

20th century

coda

the blue day

if violence is the off-guard

of time—

began at 4 a.m. yet sticks still

the milky

wishes to be dumped

a phantom underground

creek at about the juncture

near where i began

not that you died

is violent

measure

for counting today

in my throat like

cloud of a feeling that

from a new-condo crane onto

river— maybe garrison

where bloor meets christie

my tenure as a ghost

but i wanted to—

what   kind

swings   at my

intelligence   you burn   in

of a   city   without

breathe?

in the   rush hour   and people   say

balloon   caught

inconvenient for

that large   swaying

of a   sad   carcass   somebody

of a   grievance

ankles   like the

artificial

a fireplace   in the midst

enough   trees   left to

i   leave   myself   dangling

look   at   that dark

on the crane-hook   how

the operator—

purple   bag   in the   shape

please   cut it   loose—

they bellow

wearing a pop song as a

swallow or spit?

of a violent time

offices stay up and ready

but i am no future-ghost—

strung up by the knees on

ossington while everyone heads

a blue sky—

in this poem

hoisted—

below   like they're

mouthguard   —what to   do—

each note   is the   interval

now that   bright-lit

despite the   perpetual   flooding

only a   comical   gif

a crane-hook   asway over

for   lattes   —such

such a   dark   body-shape

entirely   made up and

gloomily swooping

waterway— underneath us

persisting—

toward the huge lake—

toward the moon—

for all the burials—

attractive to the modern

reroute the natural

like a large force-fed bat

above the buried

the network of rivers

their pull

circulatory yank

the tug of grievance

why was it so

city to stymy and

flow and crest of creeks

and streams?

replacing blood —the fiction

who cares if

there's a nest

the present

is where

our red chorus

harness

a compromise— haunches shaking and

something about money

of financial superiority—

in forty years

egg but no nest?

we're droning

rightside down in a flipped

trying to prettily hum as

charged with the lives yet

to come

ACKNOWLEDGMENTS

This poem cycle found its gutter-leaping energies in notebook composition by hand beginning in 2014. Some of the poems reference Sudbury, my hometown, site of Ramsey Lake, hospital, hurricane, and Science North. I am indebted to two writing residencies which cultivated this work: first, at Western University in 2016–2017, where I thank Kathryn Mockler, Christopher Keep, Thy Phu, and other colleagues, as well as Madeline and Thomas Lennon for a three-month stay in their beautiful meadow-facing home; and second, at the University of Alberta in 2017–2018, where I warmly thank Janice Williamson and Marilyn Dumont as well as many new friends and colleagues, and Sheila Greckol, whose hospitality over the fall of 2017 was extraordinary and deeply appreciated. Thank you as well to the Banff Centre Leighton Colony for a retreat in November 2017 during which this manuscript stirred toward book form. Abiding thanks to long-time friends Sonja Greckol, Athina Goldberg, Victoria Freeman, Mark Fawcett, Cynthia Leroy, and many other writing pals, and a special thanks to my cherished siblings. ZCS, you are breathing itself.

"Charger 1" was published on the occasion of HIJ Reading Series No. 18 in March 2016 – thanks to Jay and Hazel Millar – and later included in *Space Between Her Lips*, thanks to Gregory Betts. I thank Sina Queyras and *Lemon Hound* 3.0 for publishing "charger 10" in fall 2017. This poem makes reference to women's reproductive rights as argued by the Alabama abortionist Willie Parker; find him online in conversation with the CBC program *Tapestry*.

Thank you to the fine people at Talonbooks, especially Catriona Strang, for her editorial patience and caring attention.

MARGARET CHRISTAKOS is attached to this earth. Born and raised in Sudbury, Ontario, she has worked as a poet, writer, editor, instructor, and poetry-culture builder in Toronto since the late 1980s. Her body of work includes nine collections of poetry, numerous chapbooks, a novel, and an inter-genre memoir. She has been shortlisted for the Trillium Book Award and the Pat Lowther Memorial Award, and is a recipient of the ReLit Award for poetry and the Bliss Carman Award. *Space Between Her Lips: The Poetry of Margaret Christakos* was published in 2017 (Laurier Poetry series). She has held appointments as Writer in Residence at the University of Windsor, Western University, London Public Library, and the University of Alberta. She is associate faculty with the MFA program in creative writing at University of Guelph-Humber and has taught widely as a sessional, most recently at Ryerson University. In 2018–2019, she was Barker Fairley Distinguished Visitor at University College, University of Toronto. She has three adult children and lives in Toronto.